WHO LIVES IN A LAKE?

by Jenny Fretland VanVoorst

TABLE OF CONTENTS

tadpole
books

fish

Fish live in a lake.

clam

Clams live in a lake.

Crayfish live in a lake.

Snails live in a lake.

Newts live in a lake.

turtle

Turtles live in a lake.

Frogs live in a lake.

WORDS TO KNOW

clams

crayfish

fish

frogs

newts

turtles

INDEX